Sathiyamoorthy Manickkam

α-amylase enzyme production by Bacillus Lichenoformis

Sathiyamoorthy Manickkam

# α-amylase enzyme production by Bacillus Lichenoformis

LAP LAMBERT Academic Publishing

**Impressum / Imprint**

Bibliografische Information der Deutschen Nationalbibliothek: Die Deutsche Nationalbibliothek verzeichnet diese Publikation in der Deutschen Nationalbibliografie; detaillierte bibliografische Daten sind im Internet über http://dnb.d-nb.de abrufbar.

Alle in diesem Buch genannten Marken und Produktnamen unterliegen warenzeichen-, marken- oder patentrechtlichem Schutz bzw. sind Warenzeichen oder eingetragene Warenzeichen der jeweiligen Inhaber. Die Wiedergabe von Marken, Produktnamen, Gebrauchsnamen, Handelsnamen, Warenbezeichnungen u.s.w. in diesem Werk berechtigt auch ohne besondere Kennzeichnung nicht zu der Annahme, dass solche Namen im Sinne der Warenzeichen- und Markenschutzgesetzgebung als frei zu betrachten wären und daher von jedermann benutzt werden dürften.

Bibliographic information published by the Deutsche Nationalbibliothek: The Deutsche Nationalbibliothek lists this publication in the Deutsche Nationalbibliografie; detailed bibliographic data are available in the Internet at http://dnb.d-nb.de.

Any brand names and product names mentioned in this book are subject to trademark, brand or patent protection and are trademarks or registered trademarks of their respective holders. The use of brand names, product names, common names, trade names, product descriptions etc. even without a particular marking in this work is in no way to be construed to mean that such names may be regarded as unrestricted in respect of trademark and brand protection legislation and could thus be used by anyone.

Coverbild / Cover image: www.ingimage.com

Verlag / Publisher:
LAP LAMBERT Academic Publishing
ist ein Imprint der / is a trademark of
OmniScriptum GmbH & Co. KG
Heinrich-Böcking-Str. 6-8, 66121 Saarbrücken, Deutschland / Germany
Email: info@lap-publishing.com

Herstellung: siehe letzte Seite /
Printed at: see last page
**ISBN: 978-3-659-78650-1**

Zugl. / Approved by: Arunachal Pradesh, Himalayan University, 2015

Copyright © 2015 OmniScriptum GmbH & Co. KG
Alle Rechte vorbehalten. / All rights reserved. Saarbrücken 2015

## Table of Contents

Preface .................................................................................................................. 3
ACKNOWLEDGEMENTS .................................................................................. 5
ABSTRACT .......................................................................................................... 7
CHAPTER 1 .......................................................................................................... 9
  INTRODUCTION .............................................................................................. 9
    Primary structure ........................................................................................... 15
    Secondary structure ....................................................................................... 16
    Tertiary structure ........................................................................................... 16
    1.2. Intracellular Enzymes ............................................................................ 17
    1.3. Extracellular enzymes ............................................................................ 18
    1.4. Animal tissue enzymes .......................................................................... 19
    1.6. Microbial enzymes ................................................................................ 19
    1.7. Fermentation ......................................................................................... 21
    1.8. Uses of α – amylase .............................................................................. 27
    Starch conversion .......................................................................................... 27
    Detergent industry ......................................................................................... 28
    Fuel alcohol production ................................................................................. 28
    Food industry ................................................................................................. 29
    Textile industry ............................................................................................. 30
    Paper industry ............................................................................................... 30
CHAPTER 2 ........................................................................................................ 33
  LITERATURE REVIEW ................................................................................. 33
    2.1. High Temperature Alkaline α – Amylase from Bacillus Lichenoformis ... 33
    2.2. Production of α – Amylase in a low cost medium ................................ 33
    2.3. Production of α – amylase from Bacillus firmus .................................. 33
    2.4. Alkaline protease production by Bacillus lichenoformis ...................... 34
    2.5. Purification of α – amylase ................................................................... 34
    2.6. $CO_2$ effect on the formation of α – amylase ......................................... 34
    2.7. Starch Hydrolysis using α – amylase ................................................... 34
CHAPTER 3 ........................................................................................................ 35
  OBJECTIVE OF THE STUDY ....................................................................... 35
CHAPTER 4 ........................................................................................................ 37
  EXPERIMENTAL SETUP .............................................................................. 37
    4.1. Biostat E Fermentor .............................................................................. 37
    4.2. Basic Drives ......................................................................................... 38
    4.3. Dimensions of the fermentor ................................................................ 39
CHAPTER 5 ........................................................................................................ 41

| MATERIALS & METHODS | 41 |
|---|---|
| 5.1. Microbial Strain | 41 |
| 5.2. Chemicals | 41 |
| 5.3. Medium | 41 |
| 5.4. Procedure | 42 |
| 5.5. Methods | 43 |
| 5.6. Pre inoculum | 43 |
| 5.7. Biomass | 43 |
| 5.8. Enzyme Activity | 43 |
| CHAPTER 6 | 45 |
| ANALYTICAL METHODS | 45 |
| 6.1. Biomass Estimation | 45 |
| 6.2. Model Calculation | 45 |
| 6.3. Enzyme Activity Determination | 46 |
| CHAPTER 7 | 49 |
| RESULTS & DISCUSSION | 49 |
| 7.1. Production of Enzyme | 49 |
| 7.2. Effect of Corn Starch Concentration | 49 |
| 7.3. Effect of pH | 49 |
| 7.4. Effect of Temperature | 50 |
| 7.5. α – amylase production in low cost medium | 50 |
| CONCLUSION | 69 |
| BIBLIOGRAPHY | 70 |

# Preface

*It gives me a great pleasure to present the book "α-amylase enzyme production by Bacillus Lichenoformis" to students and researchers in the field of Biotechnology and biochemical engineering.*

*The prime objective of this book is to present the fundamentals of enzymes and its production methods using low cost medium, instead of using conventional high cost medium. The subject matter is presented in a simple and lucid language and fairly good explanations are given for each chapters. Each chapters are thoroughly checked to make the content error free.*

*Enzymes are giant macromolecules which catalyze biochemical reactions. They are remarkable in many ways. Their three-dimensional structures are highly complex, yet they are formed by spontaneous folding of a linear polypeptide chain. Their catalytic properties are far more impressive than synthetic catalysts which operate under more extreme conditions. Each enzyme catalysis a single chemical reaction on a particular chemical substrate with very high selectivity and specificity at rates which approach "catalytic perfection". Living cells are capable of carrying out a huge repertoire of enzyme-catalyzed chemical reactions, some of which have little or no precedent in organic chemistry.*

*I am very thankful to my family, friends and staff members for appreciating my book. I welcome suggestions from students and researchers for betterment of the book.*

*I hope and believe that researchers and students will find this book very simple and useful.*

*I would welcome and appreciate comments, constructive criticism and suggestions from students and researchers for further improving the quality of the book.*

**M. Sathiyamoorthy,**
Faculty- Chemical Engg,
Higher Colleges of Tech,
Ruwais, Abu Dhabi,
United Arab Emirates.

# ACKNOWLEDGEMENTS

I express my sincere thanks and deep sense of gratitude to my Ph.D. supervisor **Dr.S.THENESHKUMAR,** Assistant Professor (Senior Grade), Department of Chemical Engineering, SRM University, Chennai, India, for his excellent guidance, active cooperation and encouragement that had been instrumental for the successful completion of this experimental research work. I would like to thank him for encouraging my research and for allowing me to grow as a research person. His advice on both research as well as on my career have been invaluable. His valuable suggestions in bringing out the present investigation to the ultimate form of this Ph.D. thesis, are difficult to express in words.

I take this opportunity to convey my sincere thanks to **Dr.K.SRINIVASAN,** for being there for me throughout the entire doctorate program and his excitement and willingness to provide feedback made the completion of this research an enjoyable experience.

I am thankful to **Dr.R.MATHAVAN,** for his support and guidance made my thesis work possible. He has been actively interested in my work and has been available to advise me. I am very grateful for his patience, motivation, enthusiasm and immense knowledge, make him a great organizer.

A special thanks to my family. Words cannot express how grateful I am to my loving wife **Mrs.S.KOUSALYA,** Mother, Mother in law, Father in Law, My sweet daughters Pooja & Harini that you have made on my behalf. Yours prayers for me was what sustained me thus for.

Thanks are due to all the staff of Chemical Engineering Department, Himalayan University for extending their cooperation in successfully completing this research work in time.

# ABSTRACT

Production of α-amylase enzyme by *Bacillus Lichenoformis* using stirred tank fermenter (BIOSTAT – E) was carried out. The strain was obtained from National Chemical Laboratory, Pune, India. Corn starch is used as the substrate. The enzyme production was studied by changing the various parameters like temperature, pH, rpm and substrate concentration. The enzyme activity shows maximum at a temperature of $35^0C - 37^0C$, pH 8 and 300 rpm. The maximum enzyme production was achieved, for 1% concentration of cornstarch at $35^0C$ and pH 8 using the fermentation medium contains yeast extract and peptone and the enzyme activity was found to be 45 DUN/ml (Dextrinizing unit/ml). Since the cost of yeast extract and peptone is very high, so the further work was done using some low cost carbon and nitrogen sources like defatted cotton seed, defatted soya flour and mustard seed which are extracted from agricultural byproducts. The enzyme activity for using the low cost medium was found to be nearly triple such as 121 DUN/ml. The enzyme production reaches the steady phase at 24 hours. So it is highly recommended that using the low cost medium for the α-amylase enzyme gives better enzyme activity.

# CHAPTER 1.

## INTRODUCTION

### 1.1. History of Enzymes & its Classification

Enzymes are Proteins which catalyze variety of reaction in the Biological systems. When enzymes were first intensively studied in the last two centuries this Chemical Nature was obscure and even the reactions catalyzed were frequently ill defined. It was natural and therefore, that individual enzymes were given names by their discoverers. Most enzymes are studied and need to be named before any significant information about their structures exists. Wherever the 'same' enzyme from different organism is studied, it is found that Proteins different in detailed structure (and some times in gross structure) can have essentially the same catalytic properties. In the recommendations of the "International Union of Biochemistry Nomenclature Committee (1984), therefore, an enzyme name does not specify a structure but instead defines the Principal reaction catalyzed.

The use of enzymes in the diagnosis of disease is one of the important benefits derived from the intensive research in biochemistry since the 1940's. Enzymes have provided the basis for the field of clinical chemistry. It is, however, only within the recent past few decades that interest in diagnostic enzymology has multiplied. Many methods currently on record in the literature are not in wide use, and there are still large areas of medical research in which the diagnostic potential of enzyme reactions has not been explored at all.

This section has been prepared by Worthington Biochemical Corporation as a practical introduction to enzymology. Because of its close involvement over the years in the theoretical as well as the practical aspects of enzymology,

Worthington's knowledge covers a broad spectrum of the subject. Some of this information has been assembled here for the benefit of laboratory personnel.

The living cell is the site of tremendous biochemical activity called metabolism. This is the process of chemical and physical change which goes on continually in the living organism. Build-up of new tissue, replacement of old tissue, conversion of food to energy, disposal of waste materials, reproduction - all the activities that we characterize as "life."

This building up and tearing down takes place in the face of an apparent paradox. The greatest majority of these biochemical reactions do not take place spontaneously. The phenomenon of catalysis makes possible biochemical reactions necessary for all life processes. Catalysis is defined as the acceleration of a chemical reaction by some substance which itself undergoes no permanent chemical change. The catalysts of biochemical reactions are enzymes and are responsible for bringing about almost all of the chemical reactions in living organisms. Without enzymes, these reactions take place at a rate far too slow for the pace of metabolism.

The oxidation of a fatty acid to carbon dioxide and water is not a gentle process in a test tube - extremes of pH, high temperatures and corrosive chemicals are required. Yet in the body, such a reaction takes place smoothly and rapidly within a narrow range of pH and temperature. In the laboratory, the average protein must be boiled for about 24 hours in a 20% HCl solution to achieve a complete breakdown. In the body, the breakdown takes place in four hours or less under conditions of mild physiological temperature and pH.

The existence of enzymes has been known for well over a century. Some of the earliest studies were performed in 1835 by the Swedish chemist Jon Jakob Berzelius who termed their chemical action catalytic. It was not until 1926, however, that the first enzyme was obtained in pure form, a feat accomplished by James B. Sumner of Cornell University. Sumner was able to isolate and crystallize the enzyme urease from the jack bean. His work was to earn him the 1947 Nobel Prize. John H. Northrop and Wendell M. Stanley of the Rockefeller Institute for Medical Research shared the 1947 Nobel Prize with Sumner. They discovered a complex procedure for isolating pepsin. This precipitation technique devised by Northrop and Stanley has been used to crystallize several enzymes.

All known enzymes are proteins. They are high molecular weight compounds made up principally of chains of amino acids linked together by peptide bonds. Enzymes can be denatured and precipitated with salts, solvents and other reagents. They have molecular weights ranging from 10,000 to 2,000,000. Many enzymes require the presence of other compounds - cofactors - before their catalytic activity can be exerted. This entire active complex is referred to as the holoenzyme; i.e., apoenzyme (protein portion) plus the cofactor (coenzyme, prosthetic group or metal-ion-activator) is called the holoenzyme.

**Apo enzyme + Cofactor = Holoenzyme**

According to Holum, the cofactor may be:

*1. A coenzyme - a non-protein organic substance which is dialyzable, thermostable and loosely attached to the protein part.*

*2. A prosthetic group - an organic substance which is dialyzable and thermostable which is firmly attached to the protein or Apo enzyme portion.*

*3. A metal-ion-activator - these include $K^+$, $Fe^{++}$, $Fe^{+++}$, $Cu^{++}$, $Co^{++}$, $Zn^{++}$, $Mn^{++}$, $Mg^{++}$, $Ca^{++}$, and $Mo^{+++}$.*

Except for some of the originally studied enzymes such as pepsin, rennin, and trypsin, most enzyme names end in "ase". The International Union of Biochemistry (I.U.B.) initiated standards of enzyme nomenclature which recommend that enzyme names indicate both the substrate acted upon and the type of reaction catalyzed. Under this system, the enzyme uricase is called urate: $O_2$

oxidoreductase, while the enzyme glutamic oxaloacetic transaminase (GOT) is called L-aspartate: 2-oxoglutarate aminotransferase.

Enzymes are classified in to six classes. Enzymes in the first three classes all catalyze transfer reactions, with stoichiometry A+B → P+Q, but differ in other respects. Oxidoreductases catalyze reaction in which one or more electronics (usually two) are transferred from a donor (reducing agent) to an acceptor (Oxidizing agent). In many oxidoreductases the oxidized substrate can be regarded as a hydrogen donor, and for these enzymes the term dehydrogenase is preferred. Hydrolases catalyze hydrolytic reaction, i.e. reactions in which water is the acceptor of the transferred group. The transferases thus comprise all enzymes catalyzing transfer reaction that are not oxide reductases or hydrolases.

Lyases catalyze elimination reaction, where the bond is broken without oxidoreduction or hydrolysis and in most cases have stoichiometry. A→ P+Q. Enzymes can be classified by the kind of chemical reaction catalyzed.

1. Addition or removal of water

    a) Hydrolases - these include esterases, carbohydrases, nucleases, deaminases, amidases, and proteases

    b) Hydrases such as fumarase, enolase, aconitase and carbonic anhydrase

2. Transfer of electrons

    a) Oxidases

    b) Dehydrogenases

3. Transfer of a radical

    a) Transglycosidases - of monosaccharides

    b) Transphosphorylases and phosphomutases - of a phosphate group

    c) Transaminases - of amino group

  d) Transmethylases - of a methyl group

  e) Transacetylases - of an acetyl group

4. Splitting or forming a C-C bond

  a) Desmolases

5. Changing geometry or structure of a molecule

  a) Isomerases

6. Joining two molecules through hydrolysis of pyrophosphate bond in ATP or other tri-phosphate

  a) Ligases

The six classes are further sub divided in to subclasses, to specify the type of reaction more fully and to indicate the reactants. All the enzymes have a property of either intra cellular or extra cellular in nature. But most of them are extra cellular in nature.

Enzymes are catalysts and increase the speed of a chemical reaction without themselves undergoing any permanent chemical change. They are neither used up in the reaction nor do they appear as reaction products. The basic enzymatic reaction can be represented as follows

Where E represents the enzyme catalyzing the reaction, S the substrate, the substance being changed, and P the product of the reaction.

  A theory to explain the catalytic action of enzymes was proposed by the Swedish chemist Savante Arrhenius in 1888. He proposed that the substrate and Enzyme formed some intermediate substance which is known as the enzyme Substrate complex. The reaction can be represented as:

    Substrate + Enzyme  ⟶  Enzyme substrate complex

If this reaction is combined with the original reaction equation [1], the following results:

$$S + E \longrightarrow ES \longrightarrow P + E$$
$$\text{Substrate} \quad \text{Enzyme} \quad \text{Enzyme substrate} \quad \text{Product} \quad \text{Enzyme}$$

The existence of an intermediate enzyme-substrate complex has been demonstrated in the laboratory, for example, using catalase and a hydrogen peroxide derivative. At Yale University, Kurt G. Stern observed spectral shifts in catalase as the reaction it catalyzed proceeded. This experimental evidence indicates that the enzyme first unites in some way with the substrate and then returns to its original form after the reaction is concluded.

The study of a large number of chemical reactions reveals that most do not go to true completion. This is likewise true of enzymatically-catalyzed reactions. This is due to the reversibility of most reactions. In general:

$$A + B \underset{k-1}{\overset{k+1}{\rightleftharpoons}} C + D$$

Where $K^{+1}$ is the forward reaction rate constant and $K^{-1}$ is the rate constant for the reverse reaction.

Applying this general relationship to enzymatic reactions allows the equation:

$$S + E \underset{k-1}{\overset{k+1}{\rightleftharpoons}} ES \underset{k-2}{\overset{k+2}{\rightleftharpoons}} P + E$$

Equilibrium, a steady state condition, is reached when the forward reaction rates equal the backward rates. This is the basic equation upon which most enzyme activity studies are based.

### **Enzyme structure**

- Enzymes are **protein** macromolecules.

- They have a defined amino acid sequence, and are typically 100-500 amino acids long.
- They have a defined three-dimensional structure.
- Enzymes are **catalysts**.
    - They act as a catalyst to a chemical or biochemical reaction.
    - They increase the speed of that reaction, typically by $10^6$-$10^{14}$ times faster than the rate of the un catalyzed reaction.
    - They are *selective* for a single substrate.
    - They are *stereospecific*, meaning the reaction produces a single product.

**Primary structure**

Enzymes are made up of **α amino acids** which are linked together via amide (peptide) bonds in a linear chain. This is the **primary structure**. The resulting amino acid chain is called a *polypeptide* or *protein*. The specific order of amino acids in the protein is encoded by the DNA sequence of the corresponding gene.

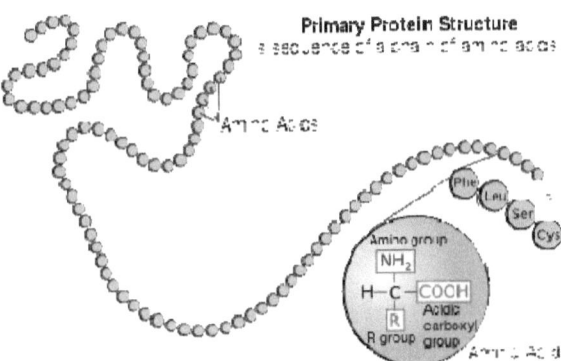

**Figure 1.1:** Primary structure of enzymes.

**Secondary structure**

Because the hydrogen in the amide group (NH$_2$) and the oxygen in the carboxyl group (COOH) of each amino acid can hydrogen bond with each other, this means that the amino acids in the same chain can interact with each other. As a result, the protein chain can fold up on itself, and it can fold up in two ways, resulting in two **secondary structures**: it can either wrap round forming the **α-helix**, or it can fold on top of itself forming the **β-sheet**.

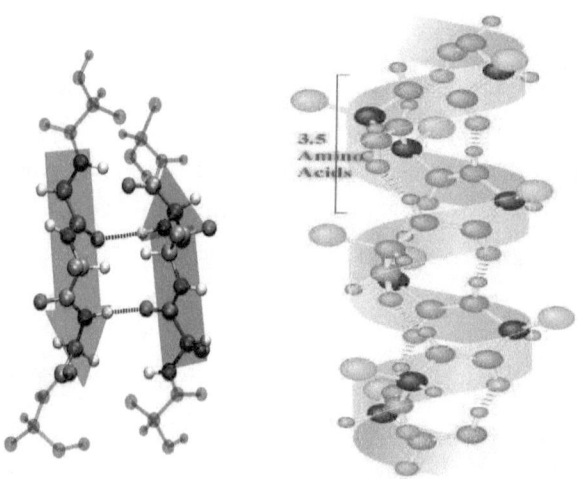

**Figure 1.2:** Secondary structure of enzymes.

**Tertiary structure**

As a consequence of the folding-up of the 2D linear chain in the secondary structure, the protein can fold up further and in doing so gains a three-dimensional structure. This is its **tertiary structure**.

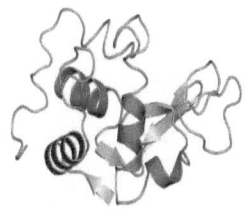

**Figure 1.3:** Tertiary structure of enzymes

## 1.2. Intracellular Enzymes

An **endoenzyme**, or **intracellular enzyme**, is an enzyme that functions within the cell in which it was produced. Because the majority of enzymes fall within this category, the term is used primarily to differentiate a specific enzyme from an exoenzyme. It is possible for a single enzyme to have both endoenzymatic and exoenzymatic functions.

Example: Glycolytic enzymes, enzymes of Kreb's Cycle. Enzymes are a type of protein that speed up chemical reactions in cells. Enzymes are specific to the job they do. Only molecules that are the correct shape can fit into the enzyme. This is called the lock and key model. Enzymes work outside of the cell (extracellular enzymes) as well as inside the cell (intracellular enzymes).

In most cases the term endoenzyme refers to an enzyme that binds to a bond 'within the body' of a large molecule - usually a polymer. For example an endoamylase would break down large amylose molecules into shorter dextrin chains. On the other hand, an exoenzyme removes subunits from the polymer one at a time from one end; in effect it can only act at the end ponts of a polymer. An exoamylase would therefore remove one glucose molecule at a time from the end of an amylose molecule.

Enzymes occur in all living cells, where they catalyze and regulate reactions of Biochemical pathways essential to the existence of the living system. In general substrates for these enzymes are small molecular weight molecules, e.g. Sugars, amino acids, carboxylic acids, which are able to permeate the membrane. Their catalytic properties are regulates by conformational changes in their three dimensional structure accomplished by allosteric cofactor molecules.

### 1.3. Extracellular enzymes

An **exoenzyme**, or **extracellular enzyme**, is an enzyme that is secreted by a cell and functions outside of that cell. Exoenzymes are produced by both prokaryotic and eukaryotic cells and have been shown to be a crucial component of many biological processes. Most often these enzymes are involved in the breakdown of larger macromolecules. The breakdown of these larger macromolecules is critical for allowing their constituents to pass through the cell membrane and enter into the cell. For humans and other complex organisms, this process is best characterized by the digestive system which breaks down solid food via exoenzymes. The small molecules, generated by the exoenzyme activity, enter into cells and are utilized for various cellular functions. Bacteria and fungi also produce exoenzymes to digest nutrients in their environment, and these organisms can be used to conduct laboratory assays to identify the presence and function of such exoenzymes.

Extra cellular enzymes were originally defined as enzymes which are external to the cell wall and in contact with surrounding medium. At present we consider transport the membrane as the primary secretion event. Thus for the purpose of this review the term & erection is used to refer to the transmembrane passage of protein and the term extra cellular to those proteins that have undergone this process. The biological function of this kind of enzymes may be seen in the hydrolysis of macro molecules which are too large to be transported in to the cell.

## 1.4. Animal tissue enzymes

Enzymes used in Industry are isolated from animal and plant tissues, as well as from Micro organisms. One of these three sources may be favored for a given enzyme. For example, some proteolytic enzymes isolated from animals may be advantageous in special fields of application. The enzyme chymosin, also known as rennet, is an acid protease used in the milk-clotting step of cheese production. A mixture of chymosin and its zymogen prochymosin, which may be converted chymosin by low pH treatment, are currently obtained from the abo-masum of an unweaned calf. Animal glands, e.g. the pancreas, are sources for hydrolyzing enzymes used as a digestive acids. The pancreas is a very rich sources of enzymes. It contains about 23% of trypsinogen and 10 -14% of chymotrypsinogen. So called pancreatin, a digestive aid, contains several enzymes such as amylase, lipase and protease.

## 1.5. Plant tissue enzymes

Plant protease isolated from pineapple (bromelain) and the papaya plant (papain) have been used for meat tenderizing and chill proofing beer. Useful amylolytic enzymes occur in plant tissues such as barely, wheat, rye, Potatoes, sweet potatoes, beans, soy beans, $\alpha$ - amylase, $\beta$ - amylase, which starts at the non reducing ends of the outer chains of the starch and proceeds by gradual removal of maltose units and debranching enzyme which hydrolyzes the $\alpha$ -1 - 6 linkages of starch, were detected in these plants.

## 1.6. Microbial enzymes

Microorganisms have become increasingly important as producers of industrial enzymes and in fact most enzymes used in industry today are of Microbial origin. Attempts are now being made to replace enzymes which traditionally have been isolated from animal tissue and plant tissues with enzymes from Microorganisms. Examples for partial replacement of plant and animal enzymes in dudes. Amylases and endo - β - glucanases of malted Barley and wheat by enzymes from Bacillus and Aspergillus in the beer, distillery, baking and textile industries. Plant and animal proteases by Aspergillus and Thermoactinomyces protease for meat tenderization and for chill proofing beer. Pancreatic protease by Aspergillus and Bacillus proteases for detergent preparations and softening leather. The production of α-amylase is essential for conversion of starches into oligosaccharides. Starch is an important constituent of the human diet and is a major storage product of many economically important crops such as wheat, rice, maize, tapioca, and potato. Starch-converting enzymes are used in the production of maltodextrin, modified starches, or glucose and fructose syrups. A large number of microbial α-amylases has applications in different industrial sectors such as food, textile, paper and detergent industries. The production of α-amylases has generally been carried out using submerged fermentation, but solid state fermentation systems appear as a promising technology. Some examples of technical enzymes from animal, plant and Microorganisms sources are listed in the following table.

**Table 1.1:** List of microbial Enzymes

| Enzyme | Isolated from |
|---|---|
| **ANIMAL ENZYMES :** | |
| α – Amylase | - Pancreatic gland |
| Rennet | - Fourth stomach of calf |
| Glucose Oxidase | - Liver |
| Pepsin | - Fundus part of hog stomach |
| **PLANT ENZYMES :** | |
| α – Amylase | - Germinated barely |
| β – Amylase | - Wheat, rye, soy beans |
| Papain | - Latex of carica papaya |
| Lipoxygenase | - Beans |
| **MICRO ORGANISMS :** | |
| α – Amylase | - *Bacillus Lichenoformis* |
| Lipase | - *Candida Raugosa* |
| Protease | - *Bacillus Subtilis* |

## 1.7. Fermentation

Fermentation is applied to both the aerobic and the anaerobic metabolic activities of microorganisms in which specific Chemical changes are brought about in an organic substrate.

This Fermentation may be,
1. Solid substrate fermentation
2. Submerged fermentation.

The fermentation processes are classified in to the above two types.

**Solid Substrate fermentation**

The essential feature of solid substrate fermentation is the growth of microorganisms on a predominantly insoluble substrate without a free liquid phase. Typically, a minimum moisture content of Ca 12% exists below which microbiological activity ceases. The moisture level in solid substrate fermentation may be between 30 - 80% and for enzyme production is typically in the region of 60%. In the solid substrate fermentation some advantages and disadvantages are there.

**Advantages of Solid substrate fermentation**
- In general very simple media are used.
- Since the low moisture conditions employed are inhibitory to the growth of most bacterial species.
- Because of moisture level is low, the volume of medium per unit weight of substrate is low.
- Inoculum stages are generally unnecessary.

**Disadvantages of Solid Substrate fermentation**
- The process is limited to fungal organisms which can best tolerate low moisture conditions.
- Measurement of process parameters is difficult because of inhomogeneity of the culture.

- Control of fermentation pH, temp, oxygen transfer and moisture content is difficult.
- Most substrate (except rice) generally need pretreatment.

**Submerged fermentation**

Submerged fermentation involves growth of a Microorganism as a suspension in a liquid medium in which the various nutrients are either dissolved or, in many cases of commercial media, suspended as particulate solids.

A typical enzyme submerged fermentation involves three main stages. They are,

- Strain maintenance and seed preparation
- Inoculum growth stages
- Fermentation stages

The main factors affecting the enzyme production in submerged fermentation are,

- Temperature
- pH
- Dissolved Oxygen

**Advantage of Enzyme production by fermentation**

Production of Microbial cells can be scaled up relatively easily to allow increased production if market demands so dictate. Microbial cells have very diverse natures and the pool of possible enzyme which can in Theory, be 'tapped' is very large indeed. Microbial cells, by contrast with higher organisms, are relatively easy to culture in a controlled environment, are highly amenable to genetic alterations, because of their relatively short generation times (as low as 30 minutes in some cases) can be rapidly 'improved' by strain development techniques. A production cycle for batch fermentation will fast between one half and ten days. In contrast, for higher organisms, a life cycle lasts typically from two month to also allow the chosen strain to be grown under near ideal and closely controlled conditions to give high and consistent productivities.

## Problems faced during Fermentation

The main problem encountered with fermentation is inhibition. Some times when a large amount of substrate is present, the enzyme catalyzed reaction is diminished by the excess substrate present. This phenomenon called "Substrate inhibition". The inhibition may also occur when a large amount of Product is formed in which case it is called "Product inhibition".

Other types of inhibition may also occur when certain Chemical species other than substrate combine with the enzyme. The Chemicals tend to alter (or) modulate the catalytic activity of the enzyme and they are called modulators or effecters.

## Stock Culture

Microorganisms that carry out new fermentations or which provide higher yields for existing fermentations are of value only if they can be stored for future use in such a fashion that their growth and productive capacities remain unaltered.

There are two types of stock culture, "Working stocks" and "Primary stocks", working stock culture are used frequently, and they must be maintained in a vigorous and uncontaminated conditions. These cultures are maintained as agar slats, agar stabs, spore preparations, or broth culture, and they are held under refrigeration.

Primary stock culture that are held in reserve for presently practical or new fermentation, for comparative purposes, for biological assays, or for possible later screening programs, Transfers from these culture are made only when a new working stock culture is required, or when the primary stock culture must be sub cultured to avoid death of cells. These stock cultures are stored at room temperature are maintained in sterile Oil, or in Agar.

## Fermentation Medium

For an every individual fermentation needs most suitable medium. All Microorganisms require water, sources of energy, carbon, nitrogen, mineral elements and possibly vitamin plus oxygen is aerobic the good medium will have the following criteria.

- ❖ It will produce the maximum yield of product or biomass per gram of substrate used.
- ❖ It will produce the maximum concentration of product or biomass.
- ❖ There will be minimum yield of undesired products.

**Medium Formulation**

Medium formulation is an essential stage in the design of successful lab experiments consider the equation based on the stoichiometry for growth and product formation.

Carbon energy source Nitrogen sources other requirements.

Cell biomass + Products + $CO_2$ + $H_2O$ + heat.

The Carbon substrate has a dual role in biosynthesis and energy generation the carbon requirement under aerobic condition may be estimated from the cellular yield coefficient which is defined as

Quantity of Cell dry mater produced
Quantity of Carbon substrate utilized

**Water**

Water is the major component of all fermentation media, and is needed in many of the ancillary services. Some of the factors which need to be considered include pH, dissolved solids and effluent contamination. The mineral content of the water is very important in brewing, mashing process and beer making process. It may even be possible to reuse all the water in the fermenter feed with appropriate adjustment of nutrient levels.

**Energy Sources**

Energy for growth comes from either the oxidation of medium components or from light. Most industrial microorganisms are chemo-organotrophs, therefore the commonest source of energy will be the Carbon source such as carbohydrates, lipids and proteins.

**Carbon Sources**

It is common practice to used carbohydrates as the carbon sources in microbial fermentation processes. The most widely available carbohydrate is starch obtained from maize grain.

It is also obtained from other cereals, Potatoes and Cassava. Starch may also be readily hydrolyzed by dilute acids and enzymes to give variety of glucose preparations some common carbon sources used for media preparation is Barley grains, sucrose, lactose, commercial vegetable oils such as Olive, Maize, Cotton seed, linseed, soya beans and some fatty acids like oleic, linoleic and linolenic acids.

**Nitrogen Sources**

Most microorganisms can utilize inorganic or organic sources of nitrogen inorganic nitrogen may be supplied as ammonia gas, ammonium salts or nitrates. Ammonium salts such as ammonium sulphate will usually produce acid condition as the $NH_4$ ion is utilized. Organic nitrogen may be supplied as amino acid, protein, and Urea. In many instances growth will be faster with a supply of organic nitrogen, and a few microorganisms have an absolute requirement for amino acids.

Some commonly used nitrogen sources for media preparations are corn steep liquor, soya meal, soy beans, pea-nut meal, cotton seed meal, Distillers soluble, casein hydrolysate, slaughter house wastes, fish meal and yeast extract. For α amylase production yeast extract and peptone was used as the nitrogen sources.

**Minerals**

In many media magnesium, phosphorus, potassium, sulphur, calcium and chlorine are essential components, and because of the concentration required, they will be added as distinct components. Others such as Cobalt, Copper, Iron, Manganese, molybdenum and zinc are also essential but are usually present as impurities in other major ingredients when synthetic media are used the minor elements will have to be deliberately added.

**Vitamin Sources**

Many of the natural carbon and nitrogen sources contain all or some of the required Vitamins. When there is a Vitamin deficiency it is often possible, by careful blending of materials to eliminate the deficiency. Some production strains may also require Thiamine.

**Nutrient Recycle**

Some large scale continuous culture fermenters for SCP are designed to operate with recycle of the liquid feed. Obviously there will be a need for appropriate adjustment of nutrient levels. For this reason, $H_3PO_4$ may be used as a reagent for flocculating bacteria.

**Buffers**

The control of pH may be extremely important if optimal productivity is to be achieved. A compound may be added to the medium to serve specifically as buffer.

**Characteristics of α – amylase**

Amylolytic enzymes are widely distributed in bacteria and fungi. Amylase hydrolyze α -1, 4 - D glucosidic linkages of amylose and amylopectin. Amylases are classified into three main groups, 1. Exo-acting, 2. endo-acting, 3. Debranching. The α – amylase with optimal activity at the highest temperature (80 – 90$^0$C) is formed by the moderate thermophile *Bacillus lichenoformis*.
The some characteristics of α – amylase are,

❖ Produces products which have the α – Configuration at carbon atom of the reducing glucose unit.

- ❖ Possess an endo – attract Mechanism.
- ❖ Rapidly decreases the ability of amylose to stain blue with iodine.
- ❖ Possess the ability to bypass α - 1→ 6 branch points.

## 1.8. Uses of α – amylase

The enzyme α –amylase is used as a biocatalyst in many small scale and large scale industries some of the uses are.

- ❖ The Bacterial α – amylase used in starch hydrolysis industries, Brewing industries, Detergents industries and textile industries.
- ❖ The fungal α –amylase used in starch industries and baking industries.
- ❖ The α – amylase from Malt used as a digestive aid and supplement to bread.
- ❖ The α – amylase from *Aspergillus Orygaze* is used to produce starch liquefying syrups.
- ❖ The α–amylase from *Bacillus Subtillis* used in Desizing textile industries, Alcohol fermentation industries and glucose producing industries.
- ❖ The α – amylase produced from *Aspergillus Niger* is highly acid resistant is used as a digestive acid at pH-5.
- ❖ The α – amylase from *Bacillus lichenoformis* is used in all starch industries and detergent industries and to produce starch sizing pastes for use in paper coatings.

## Starch conversion

The most widespread applications of α-amylases are in the starch industry, which are used for starch hydrolysis in the starch liquefaction process that converts starch into fructose and glucose syrups. The enzymatic conversion of all starch includes: gelatinization, which involves the dissolution of starch granules, thereby forming a viscous suspension; liquefaction, which involves partial hydrolysis and loss in viscosity; and saccharification, involving the production of glucose and maltose via further hydrolysis. Initially, the α-amylase of *Bacillus amyloliquefaciens* was used but it has been replaced by the α-amylase of *Bacillus stearothermophilus* or *Bacillus lichenoformis*. The enzymes from the Bacillus species are of special interest for large-scale biotechnological processes due to their remarkable thermos stability and because efficient expression systems are available for these enzymes.

## Detergent industry

Detergent industries are the primary consumers of enzymes, in terms of both volume and value. The use of enzymes in detergents formulations enhances the detergents ability to remove tough stains and making the detergent environmentally safe. Amylases are the second type of enzymes used in the formulation of enzymatic detergent, and 90% of all liquid detergents contain these enzymes. These enzymes are used in detergents for laundry and automatic dishwashing to degrade the residues of starchy foods such as potatoes, gravies, custard, chocolate, etc. to dextrins and other smaller oligosaccharides. Amylases have activity at lower temperatures and alkaline pH, maintaining the necessary stability under detergent conditions and the oxidative stability of amylases is one of the most important criteria for their use in detergents where the washing environment is very oxidizing. Removal of starch from surfaces is also important in providing a whiteness benefit, since starch can be an attractant for many types of particulate soils. Examples of amylases used in the detergent industry are derived from *Bacillus* or *Aspergillus*.

## Fuel alcohol production

Ethanol is the most utilized liquid biofuel. For the ethanol production, starch is the most used substrate due to its low price and easily available raw material in most regions of the world. In this production, starch has to be solubilized and then submitted to two enzymatic steps in order to obtain fermentable sugars. The bioconversion of starch into ethanol involves liquefaction and saccharification, where starch is converted into sugar using an amylolytic microorganism or enzymes such as α-amylase, followed by fermentation, where sugar is converted into ethanol using an ethanol fermenting microorganism such as yeast *Saccharomyces cerevisiae*. The production of ethanol by yeast fermentation plays an important role in the economy of Brazil. In order to obtain a new yeast strain that can directly produce ethanol from starch without the need for a separate saccharifying process, protoplast fusion was performed between the amylolytic yeast *Saccharomyces fibuligera* and *S. cerevisiae*. Among bacteria, α-amylase obtained from thermo resistant bacteria like *Bacillus licheniformis* or from engineered strains of *Escherichia coli* or *Bacillus subtilis* is used during the first step of hydrolysis of starch suspensions.

## Food industry

Amylases are extensively employed in processed-food industry such as baking, brewing, preparation of digestive aids, production of cakes, fruit juices and starch syrups. The α-amylases have been widely used in the baking industry. These enzymes can be added to the dough of bread to degrade the starch in the flour into smaller dextrins, which are subsequently fermented by the yeast. The addition of α-amylase to the dough results in enhancing the rate of fermentation and the reduction of the viscosity of dough, resulting in improvements in the volume and texture of the product. Moreover, it generates additional sugar in the dough, which improves the taste, crust color and toasting qualities of the bread. Besides generating fermentable compounds, α-amylases also have an anti-staling effect in bread baking, and they improve the softness retention of baked goods, increasing the shelf life of these products. Currently, a thermostable maltogenic amylase of *Bacillus stearothermophilus* is used commercially in the bakery industry. Amylases are also used for the clarification of beer or fruit juices, or for the pretreatment of animal feed to improve the digestibility of fiber.

## Textile industry

Amylases are used in textile industry for desizing process. Sizing agents like starch are applied to yarn before fabric production to ensure a fast and secure weaving process. Starch is a very attractive size, because it is cheap, easily available in most regions of the world, and it can be removed quite easily. Starch is later removed from the woven fabric in a wet-process in the textile finishing industry. Desizing involves the removal of starch from the fabric which serves as the strengthening agent to prevent breaking of the warp thread during the weaving process. The α-amylases remove selectively the size and do not attack the fibers. Amylase from *Bacillus* stain was employed in textile industries for quite a long time. Modern production processes in the textile industry can cause breaking of the warp thread. To strengthen the thread, sizing agents are used which strengthen the thread by forming a layer on it and can be removed after the fabric is woven. Starch is a preferred sizing agent as it is easily available, cheaper and can be easily removed from the fabric. The layer of starch is subjected to hydrolysis in the desizing process where α-Amylase is employed to cleave starch particles randomly into water soluble components that can be removed by washing. The enzyme acts specifically on the starch molecules alone, leaving the fibers unaffected.

## Paper industry

The use of α-amylases in the pulp and paper industry is for the modification of starch of coated paper, i.e. for the production of low-viscosity, high molecular weight starch. The coating treatment serves to make the surface of paper sufficiently smooth and strong, to improve the writing quality of the paper. In this application, the viscosity of the natural starch is too high for paper sizing and this can be altered by partially degrading the polymer with α-amylases in a batch or continuous processes. Starch is a good sizing agent for the finishing of paper, improving the quality and erase ability, besides being a good coating for the paper.

The size enhances the stiffness and strength in paper. Examples of amylases obtained from microorganisms used in paper industries.

And also used in liquefaction of heavy starch pastes which from during heating steps in the manufacture of corn and chocolate syrups.

This α – amylase used in bread production industries and removal of food spots in the dry cleaning industries.

The α – amylase from the above strain used in Textile industries for desizing of fabrics, mashing in distillery industries, production of corn syrup, sugar recovery from scarp candy and baking and mixing industries.

# CHAPTER 2.

## LITERATURE REVIEW

### 2.1. High Temperature Alkaline α – Amylase from Bacillus Lichenoformis

Mr. Pratima Bajpai and Pramod K. Bajpai were produced High temperature alkaline α - amylase from the Bacterial strain *Bacillus Lichenoformis*. They were obtained. The enzyme produced was quit active even at 100°C, and it showed the optimum activity at 90°C. It exhibited optimum activity in the broad pH range 5.5 - 10. The effect of $Na^+$ and $Ca^{2+}$ ions on enzyme activity was also studied. The fermentation was carried out in shake flask and 2.6 L Marubishi fermenter.

### 2.2. Production of α – Amylase in a low cost medium:

Mr. Pratima Bajpai and Umender Sharma were produced α -amylase in a low cost medium by the Bacterial strain *Bacillus lichenoformis*. They were used medium which contains low cost defaulted cotton seed, Defaulted Soya bean, corn steep liquor instead of yeast extract and peptone. The α – amylase enzyme of the strain showed excellent stability at high temperature and over wide pH range. This low cost medium produced 5 times more enzyme than the high cost medium which contains yeast extract and peptone in shake Flask. In a 2.6-L fermentor the enzyme production was further doubled.

### 2.3. Production of α – amylase from Bacillus firmus

Mr. Seung – Hyeon Moon and Satish J. Parulekar (8) were studied the production of α – amylase using the Bacterial species *Bacillus firmus* in Batch. Fed batch and continuous suspension cultures. They optimized the fermentation conditions like temperature, pH and aeration etc.

## 2.4. Alkaline protease production by Bacillus lichenoformis

Mr. Young hee lee and History of Nam chang were produced alkaline protease using the same Bacterial species *Bacillus Lichenoformis* in an aqueous two phase system which contains 5% w/w dextran and 5% w/w polyethylene glycol 6000. The two – phase system produced less enzyme in total amount than the control in the early phase, but after 50 hours the enzyme activity reached 1.5 times more than the control culture.

## 2.5. Purification of α – amylase

Mr. Martha H. Moseley and Leonard Keay were got the crude *Bacillus subtillis* and produced α – amylase. The purpose of this research was to develop a purification procedure for the α – amylase and to characterize the product. Several purification procedures are present in this literature. Characterization of other Subtillis α – amylase can also be found in the literature.

## 2.6. $CO_2$ effect on the formation of α – amylase

Mr. A.P. Gandhi and L. Kjaergaard were studied the effect of carbon dioxide on the formation of α – amylase by *Bacillus subtilis* is growing in continuous and Batch cultures. Different levels of $CO_2$ examined in the batch cultures stimulated the formation of α – amylase, with the highest activity being obtained using 6% $CO_2$ (v/v). In chemostat cultures, air containing 8% $CO_2$ increased the sp.enzyme productivity almost three times over the control, without affecting the cell growth.

## 2.7. Starch Hydrolysis using α – amylase

Mr. Yi-Hsuju, Wen-Jen Chen and Cheng-Kang Lee (12) were studied the starch slurry hydrolysis using α – amylase immobilized on a hollow-fiber reactor. Starch slurry of 30% (w/v) can be hydrolyzed without any difficulties by the hollow-fiber reactor.

# CHAPTER 3.

## OBJECTIVE OF THE STUDY

Enzymes are Proteins which catalyze variety of reaction in the Biological systems. There are many methods used to produce the enzymes among that the biological methods are widely used. In this type of biological method of production, solid state fermentation is applied for the production. In all the types of fermentation processes, the cultures has been prepared using yeast extract and peptone etc. These are added to the culture in terms of nutrients as a carbon and nitrogen sources for the microorganism. The cost of these chemicals are much expensive. So the alternative method has been proposed for the preparation of culture medium using some low cost agricultural byproducts such as defatted cotton seed, defatted soybean, mustard seed etc. The fermentation has to carryout using these type of low cost medium to check the productivity and enzyme activity.

# CHAPTER 4.

## EXPERIMENTAL SETUP

### 4.1. Biostat E Fermentor

The fermentation was carried out in a B. BRAUN CO, Biostat E fermentor. It is a compact and comprehensive fermentation system on a laboratory scale, which can be used in microbiological and biotechnological research and development. Biostat E fermentors are designed for use in discontinuous fermentation (Batch operations) as well as in continuous process. The measurement and control system used in compatible with computers. The Biostat E is protected against unauthorized use with a main key. All modules of the measurement and control section are separately switched on. Therefore they can be installed independently from the control in spite of the central mains switch. Additional modules can be inserted without interruption or disturbance of operations.

**Figure 4.1:** Biostat E Fermentor

## 4.2. Basic Drives

The lower front panel of the basic device is provided with installation ports for at least 4 dosing pumps of the four, three are peristaltic pumps for the supply of acid, alkali and antifoam agent, the fourth is prepared to install precision dosing pumps.

The arrangements of the various technical appliances in the basic devices are:

- Thermostat system which containing heating and cooling water circuit for tempering as well as for sterilization.
- Gas supply system including exhaust equipment.
- Motor and drive system for the stirrer shaft drive.

The recorder, of 6 channels dot printer records the following measurement values in the basic devices.

- Temperature
- Speed
- pH Value
- Antifoam consumption

The culture vessel is mounted on the console laterally fixed at the fermentor where there are the corresponding borings for the feet of the culture. Simultaneously the connection to the stirrer drive is guaranteed. For starting operating the device the filling state of the fermentor thermostat is to be checked. The set point temperature is adjusted at the corresponding digital switch of the module. A good mixing of the culture vessel is a prerequisite. For that a stirrer system is provided which is driven by a controlled DC motor. The stirrer speech can be directly adjusted by the digital switch of the speed controlled module. The adjustable speed range is 50 – 1500 minutes$^{-1}$.

The pH – value in the culture medium can be electro chemically determined via a combined – glass electrode. The pH set point desired can be adjusted with the digital switch of the pH controller.

## 4.3. Dimensions of the fermentor

| | |
|---|---|
| Total volume of the fermentor : | 8 lit. |
| Working volume : | 7 lit. |
| Max working temperature : | 138°C |
| Max working pressure : | 12 atm |
| Diameter of the fermentor : | 17.5 cm |
| Height of the fermentor : | 40 cm |
| Agitator type : | 6 Blade, Paddle type Agitator |

# CHAPTER 5.

## MATERIALS & METHODS

## 5.1. Microbial Strain

*Bacillus Lichenoformis* NCIM 2051 Received from National Chemical Laboratory, Pune, India.

## 5.2. Chemicals

| | | |
|---|---|---|
| Beef extract | : | Loba Chemie Pvt. Ltd. Mumbai |
| Peptone | : | Loba Chemie Pvt. Ltd. Mumbai |
| NaCl | : | Fischer inorganic & arome, Chennai |
| $MgSO_4$ | : | S.D. Fine Chemical, Biosar |
| $KH_2PO_4$ | : | S.D. Fine Chemical, Biosar |
| $CaCl_2$ | : | S.D. Fine Chemical, Biosar |
| Yeast extract | : | Qualigens, Mumbai |
| Agar | : | Ranbaxy Ltd,. S.A.S. Nagar |
| Corn Starch & Mustard | : | Local market |
| Defatted Cotton Seed | : | Local market |
| Defatted Soya flour | : | Loba Chemie Pvt. Ltd., Mumbai |

## 5.3. Medium

### Universal Medium for Bacteria

| | | |
|---|---|---|
| Beef extract | : | 1.0 % |
| Sodium Chloride | : | 0.5 % |
| Peptone | : | 1.0 % |
| pH | : | 7.0 - 7.2 |

Sterilize the medium, and adjust the pH at 7.2.    Add 2% Agar for making slants.

### Corn starch medium: (Basal Medium)

| | | |
|---|---|---|
| Corn starch | : | 1 % |
| Yeast extract | : | 0.2 % |
| Peptone | : | 0.5 % |
| $MgSO_4$ | : | 0.05 % |
| $KH_2PO_4$ | : | 0.05 % |
| NaCl | : | 0.15 % |
| $CaCl_2$ | : | 0.015 % |

### Low cost medium

| | | |
|---|---|---|
| Corn starch | : | 1 % |
| $MgSO_4$ | : | 0.05 % |
| $KH_2PO_4$ | : | 0.05 % |
| NaCl | : | 0.15 % |
| $CaCl_2$ | : | 0.015 % |
| Soya bean | : | 0.5 % |
| Mustard Seed | : | 2 % |
| Cotton seed | : | 3 % |

## 5.4. Procedure

Shake flask cultures were operated at constant temperature of $37^0C$ and fixed rpm with 100 ml of medium in a 500 ml Erlenmeyer flask and inoculated with the culture. Fermentation studies were carried out in above described B. Braun Biostat E fermentor with the cultural conditions of $37^0C$, pH 7, and 300 rpm. Since it is an aerobic fermentation, the aerobic rate was maintained at 1 vvm.

For every six hours the sample were collected from the sampling point provided in the top of the culture vessel, and analyzed.

## 5.5. Methods

### Stock Culture

*Bacillus Lichenoformis* NCIM 2051 was maintained in an Agar slant at $4^0C$.

### Sub Culture Maintenance

Subculture was prepared using a universal Bacteria medium and it was maintained in an incubator at $37^0C$.

## 5.6. Pre inoculum

Take 100 ml of the Universal medium inoculate this with a stock agar culture in a 500 ml Erlenmeyer flask and kept in a shaker at 300 rpm and $37^0C$. It is also called as seeding of culture.

## 5.7. Biomass

Biomass was estimated by the method of dry weight for every sample. It was expressed in terms of 1 dry wt. /lit.

## 5.8. Enzyme Activity

One unit of enzyme activity (DUN) is defined as the quantity of enzyme that causes 1 % reduction of blue color intensity of starch-iodine solution in 1 min. The optical density was first measured at 660 nm using an UV spectrophotometer.

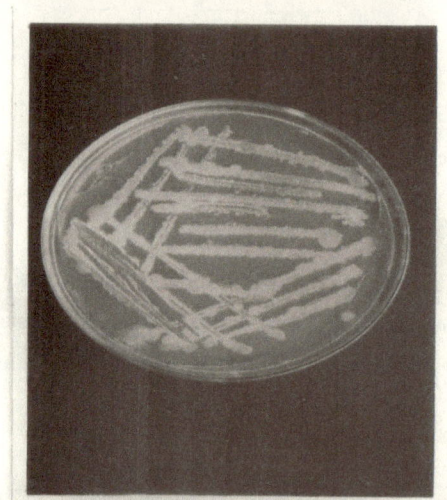

**Figure 5.1:** Culture growth in petri dish

# CHAPTER 6.

## ANALYTICAL METHODS

In this research work, the following parameter was found out by some analytical methods. They are:
1. Biomass (or) Cell concentration
2. Enzyme activity

## 6.1. Biomass Estimation

The Biomass for the sample, which got from the fermentation broth was determined by the dry weight method. Take some known amount of liquid from the fermentation of liquid from the fermentation broth in the centrifuge test tube, and kept in a centrifuge for 20 mins at 5000 rpm. The supernatant liquid was collected and kept for α – amylase activity determination. The cells settled in the bottom of the centrifuge tube was transferred to a funnel contains the gravimetric filter paper (ash less), and washed thoroughly with distilled water. Transfer this gravimetric filter paper in to the known weight silica crucible, and incinerate for 30 min. Cool the contents and measures the weight from this calculate the cell concentration.

## 6.2. Model Calculation

| | | |
|---|---|---|
| Weight of empty crucible | : | 15.8603 gms. |
| Volume of fermentation liquid taken | : | 11.4 ml |
| Weight of crucible with cells | : | 15.8895 gms. |
| Cell mass/volume of liquid taken | : | 15.8895 – 15.8603 |
| | | 0.0292 gms |
| For 1 liter | : | 2.57 |
| Cell concentration | : | 2.57 gm dry wt / lit |

## 6.3. Enzyme Activity Determination

Different techniques have been used to measure enzyme activities. There is no general method equally applicable to all enzymes. The enzyme activity may be depends on the time, enzyme concentration, substrate concentration.

Extra cellular amylase activity was determined by measuring the decrease in iodine color reaction showing dextrinization of starch. The reaction contained 1 ml of enzyme (cell free supernatant) and 10ml of 1% starch solution incubated at $40^0C$ for 10 min. The reaction was stopped by adding 10ml of 0.1N HCl. 1 ml of this acidified solution was added to 10ml 0.1N HCl. From this 1ml was added to iodine solution (0.05% iodine in 0.5% Kl). The optical density of the blue colored solution was determined of 660 nm one unit of enzyme activity (DUN) is defined as the quantity of enzyme that causes 1% reduction of blue color intensity of starch iodine solution at $40^0C$ in 1 min.

For amylase activity determination requires the standard chart for starch iodine solution. Take six test tubes in that add 0, 0.2, 0.4, 0.6, 0.8 and 1 ml of 1% starch solution respectively and add 1, 0.8, 0.6, 0.4, 0.2 and 0 ml of water reply. Add 10ml of iodine solution (0.05% $I_2$ and 0.5% Kl) in all the six text tubes. The inference is the light blue color formation. The optical density of the blue colored solution was measured at 660 nm in the UV spectrophotometer. The standard graph was drawn by plotting starch concentration Vs absorbance. From the standard graph the enzyme activity was calculated.

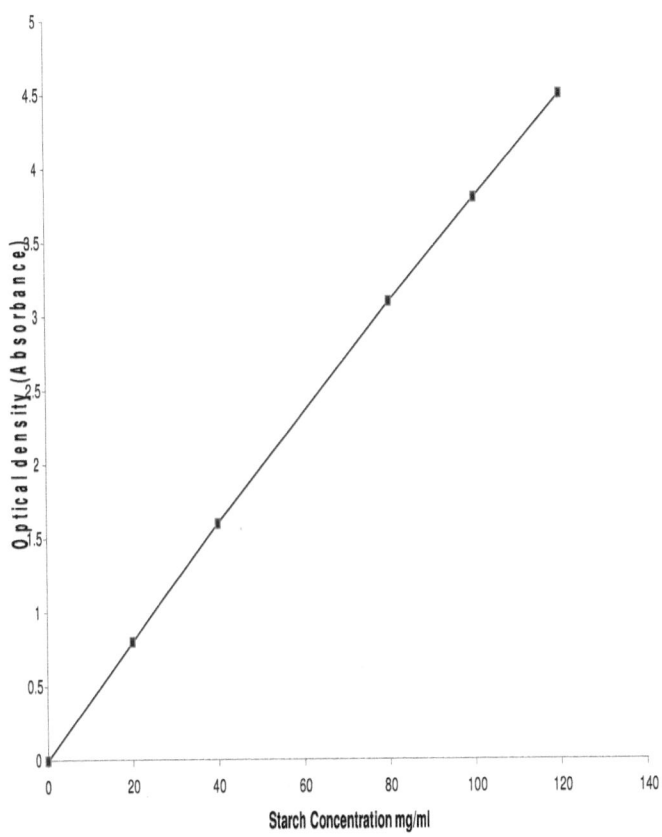

**Figure 6.1:** Calibration graph of starch iodine solution

# CHAPTER 7.

## RESULTS & DISCUSSION

### 7.1. Production of Enzyme

The growth pattern of *Bacillus Lichenoformis* NCIM 2051 and α – amylase production was observed for three days in basal medium with 1% cornstarch as a carbon source. The formation of α – amylase started from 4 hours. The maximum enzyme production was achieved at 24 hours. The pH of the broth increased from 7 at the beginning to 8.9 at the end of fermentation. The maximum yield was achieved at $35^0C$.

### 7.2. Effect of Corn Starch Concentration

The effect of corn starch concentration was further studied. The α – amylase production was studied, by changing the Corn starch concentration at 0.5%, 1% and 2.5%. It was found that with an increase of starch concentration in the medium beyond 1%, enzyme production did not increase. At higher starch concentration, enzyme production was comparatively lower and the time required to reach the maximum enzyme level was longer.

### 7.3. Effect of pH

The bacterium was found to grow at pH 3-11, with growth resulting in an increase of the patient e media's pH. Enzyme production started at 5.0 and ceased at pH 10.0. Maximum enzyme production occurred at pH 6-9. Very little enzyme production in the medium at initial pH of 3 - 4. At higher pH values (10-11), growth was quite high, but the amount of enzyme production was very low.

### 7.4. Effect of Temperature

The strain was found to grow and produce enzyme at temperatures from 25 to 50°C. Maximum enzyme production was observed at 35°C. Growth and enzyme production both started decreasing drastically above 40°C.

### 7.5. α – amylase production in low cost medium

The α – amylase production was further studied by using the low cost medium which containing the carbon and nitrogen sources like flour, mustered seeds. Since the cost of yeast extract and peptone in the Basal medium is very high, we can replace the yeast extract and peptone with the above mentioned things.

The low cost medium produced 2 times more enzyme than the high cost synthetic medium (yeast extract and peptone).the medium containing 0.5% defatted, 2% mustered seed in the place of yeast extract and peptone , was found to yield the maximum amount of α – amylase.

The experimental data are listed in the below tables for different concentration and compositions of the low cost medium.

**Batch: 1**

Table 7.1: Enzyme Production for 1% corn Starch Concentration

| S.No | Time hours | pH | Temp °C | %PO$_2$ | rpm | Cell concentration. gm dry wt/lit | Enzyme Activity DUN/ml |
|---|---|---|---|---|---|---|---|
| 1 | 0 | 7.0 | 36.9 | 104.7 | 300 | 0.052 | 0.003 |
| 2 | 3 | 7.1 | 36.8 | 101.8 | 300 | 0.236 | 2.01 |
| 3 | 6 | 7.2 | 37.0 | 100.2 | 300 | 0.480 | 3.37 |
| 4 | 12 | 5.8 | 37.1 | 22.6 | 300 | 1.215 | 4.51 |
| 5 | 18 | 7.3 | 35.7 | 89.3 | 300 | 2.882 | 29.31 |
| 6 | 24 | 8.2 | 35.6 | 86.8 | 300 | 2.843 | 55.84 |
| 7 | 48 | 9.0 | 35.6 | 96.4 | 300 | 2.745 | 55.93 |
| 8 | 72 | 8.9 | 36.2 | 98.7 | 300 | 2.461 | 56.11 |

## Batch: 2

**Table 7.2:** Enzyme Production for 2.5% corn Starch Concentration

| S.No | Time hours | pH | Temp °C | %PO$_2$ | rpm | Cell concentration. gm dry wt/lit | Enzyme Activity DUN/ml |
|---|---|---|---|---|---|---|---|
| 1 | 0 | 6.3 | 37.0 | 100.8 | 300 | 0.035 | 0.0007 |
| 2 | 3 | 5.8 | 36.4 | 100.1 | 300 | 0.21 | 1.37 |
| 3 | 6 | 6.1 | 35.9 | 92.7 | 300 | 0.53 | 2.56 |
| 4 | 12 | 6.7 | 35.7 | 95.6 | 300 | 0.78 | 8.48 |
| 5 | 18 | 6.9 | 35.4 | 97.9 | 300 | 1.52 | 33.49 |
| 6 | 24 | 7.1 | 35.3 | 98.3 | 300 | 2.61 | 40.83 |
| 7 | 48 | 7.9 | 35.1 | 88.5 | 300 | 2.48 | 40.71 |
| 8 | 72 | 8.5 | 34.8 | 84.3 | 300 | 2.57 | 40.74 |

**Batch: 3**

Table 7.3: Enzyme Production for 0.5% corn Starch Concentration

| S.No | Time hours | pH | Temp $^0C$ | %$PO_2$ | rpm | Cell concentration. gm dry wt/lit | Enzyme Activity DUN/ml |
|---|---|---|---|---|---|---|---|
| 1 | 0 | 6.1 | 37.0 | 120.3 | 300 | 0.026 | 0.0007 |
| 2 | 3 | 6.1 | 36.7 | 110.8 | 300 | 0.21 | 1.53 |
| 3 | 6 | 6.9 | 36.6 | 102.6 | 300 | 0.58 | 4.92 |
| 4 | 12 | 7.5 | 35.9 | 100.9 | 300 | 0.97 | 12.19 |
| 5 | 18 | 7.8 | 35.8 | 98.7 | 300 | 1.38 | 39.43 |
| 6 | 24 | 7.9 | 35.6 | 98.5 | 300 | 1.85 | 43.38 |
| 7 | 48 | 8.3 | 35.7 | 98.1 | 300 | 1.61 | 42.31 |
| 8 | 72 | 8.8 | 35.4 | 83.6 | 300 | 1.43 | 42.34 |

## Batch: 4

**Table 7.4:** Enzyme Production using Basal medium + 0.5% defatted Soya Flour

| S.No | Time hours | pH | Temp °C | %PO$_2$ | rpm | Cell concentration. gm dry wt/lit | Enzyme Activity DUN/ml |
|---|---|---|---|---|---|---|---|
| 1 | 0 | 7.1 | 37.0 | 98.3 | 300 | 0.042 | 0.98 |
| 2 | 3 | 7.8 | 36.8 | 90.7 | 300 | 1.19 | 11.91 |
| 3 | 6 | 8.3 | 36.1 | 83.6 | 300 | 2.25 | 27.56 |
| 4 | 12 | 8.5 | 35.7 | 85.9 | 300 | 3.37 | 48.29 |
| 5 | 18 | 9.2 | 35.4 | 70.2 | 300 | 3.93 | 61.49 |
| 6 | 24 | 9.8 | 35.3 | 56.3 | 300 | 4.89 | 81.24 |
| 7 | 48 | 9.7 | 35.4 | 79.3 | 300 | 4.91 | 81.31 |
| 8 | 72 | 9.8 | 35.3 | 73.2 | 300 | 4.87 | 80.9 |

## Batch: 5

**Table 7.5:** Enzyme Production using Basal medium + 3% defatted cotton seed

| S.No | Time hours | pH | Temp $^0$C | %PO$_2$ | rpm | Cell concentration. gm dry wt/lit | Enzyme Activity DUN/ml |
|---|---|---|---|---|---|---|---|
| 1 | 0 | 7.0 | 37.0 | 120.3 | 300 | 0.02 | 1.90 |
| 2 | 3 | 7.1 | 38.3 | 110.4 | 300 | 1.25 | 18.53 |
| 3 | 6 | 7.3 | 37.1 | 93.6 | 300 | 2.31 | 46.93 |
| 4 | 12 | 8.1 | 36.3 | 83.9 | 300 | 3.80 | 64.71 |
| 5 | 18 | 8.7 | 36.1 | 70.8 | 300 | 4.93 | 84.18 |
| 6 | 24 | 8.9 | 35.9 | 64.7 | 300 | 5.29 | 91.5 |
| 7 | 48 | 8.7 | 36.3 | 53.9 | 300 | 5.3 | 92.3 |
| 8 | 72 | 8.9 | 37.8 | 48.7 | 300 | 5.32 | 91.9 |

## Batch: 6

**Table 7.6:** Enzyme Production using Basal Medium + 2% Mustard Seed

| S.No | Time hours | pH | Temp °C | %PO$_2$ | rpm | Cell concentration. gm dry wt/lit | Enzyme Activity DUN/ml |
|---|---|---|---|---|---|---|---|
| 1 | 0 | 7.0 | 37.0 | 120.1 | 300 | 0.01 | 2.41 |
| 2 | 3 | 7.3 | 38.1 | 117.3 | 300 | 1.23 | 12.49 |
| 3 | 6 | 7.9 | 37.5 | 93.5 | 300 | 2.56 | 32.33 |
| 4 | 12 | 8.5 | 36.3 | 83.8 | 300 | 3.19 | 65.91 |
| 5 | 18 | 9.1 | 35.3 | 77.9 | 300 | 4.84 | 95.41 |
| 6 | 24 | 9.7 | 35.5 | 64.2 | 300 | 6.71 | 121.49 |
| 7 | 48 | 9.9 | 36.1 | 56.9 | 300 | 6.68 | 120.83 |
| 8 | 72 | 10.3 | 38.5 | 28.5 | 300 | 6.70 | 121.10 |

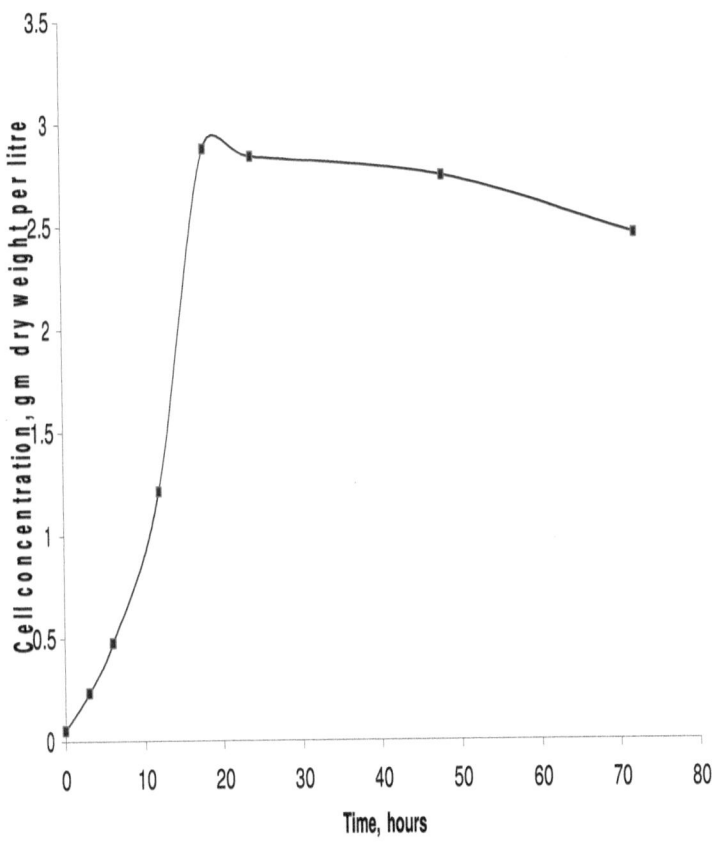

**Figure 7.1:** Biomass cell concentration for basal medium with 1% corn starch

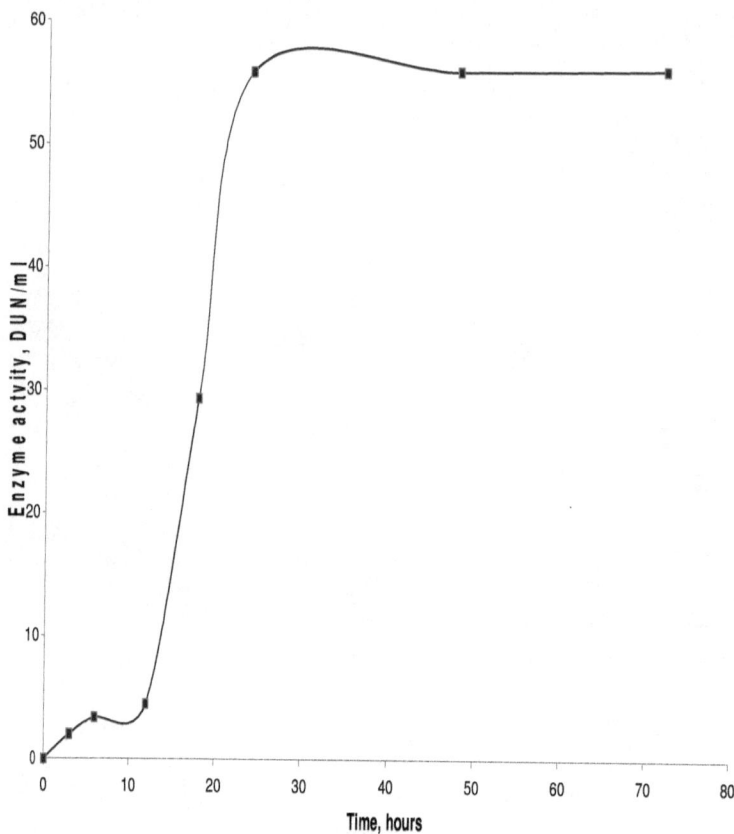

**Figure 7.2:** Enzyme activity for basal medium with 1% corn starch concentration

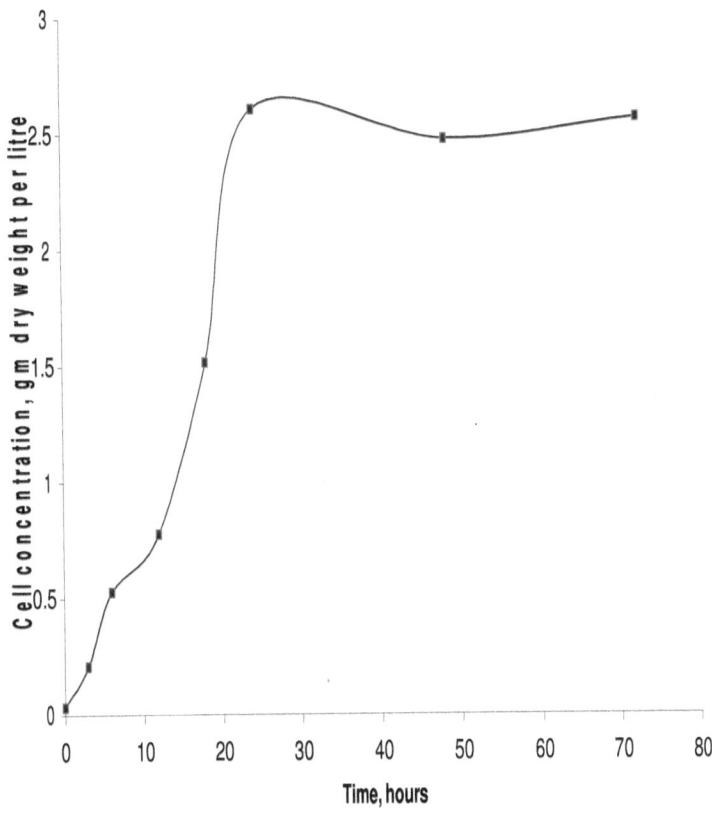

**Figure 7.3:** Biomass Cell concentration for Basal medium with 0.5% defatted Soya flour

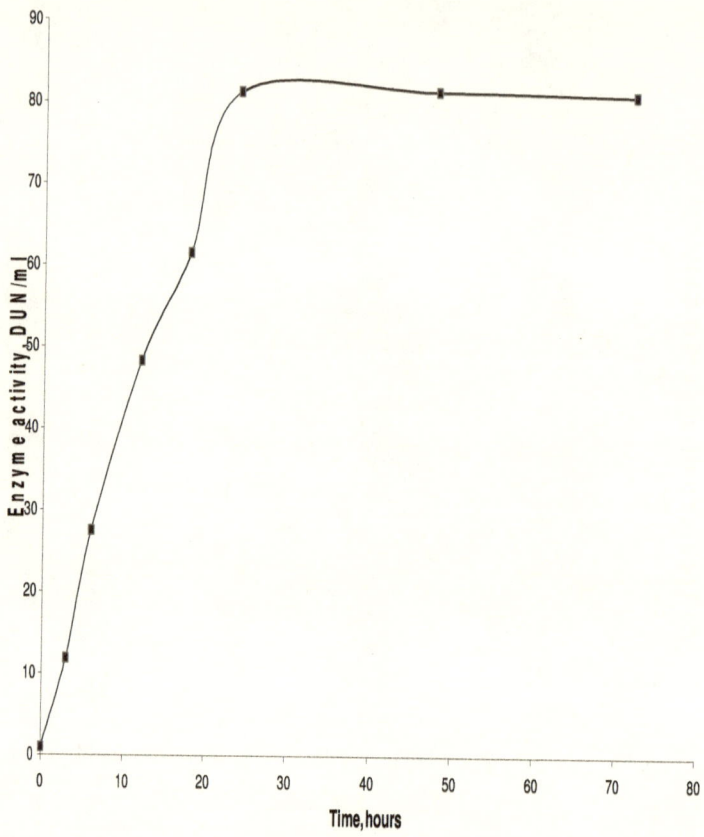

**Figure 7.4:** Enzyme activity for Basal medium with 0.5% defatted Soya flour concentration

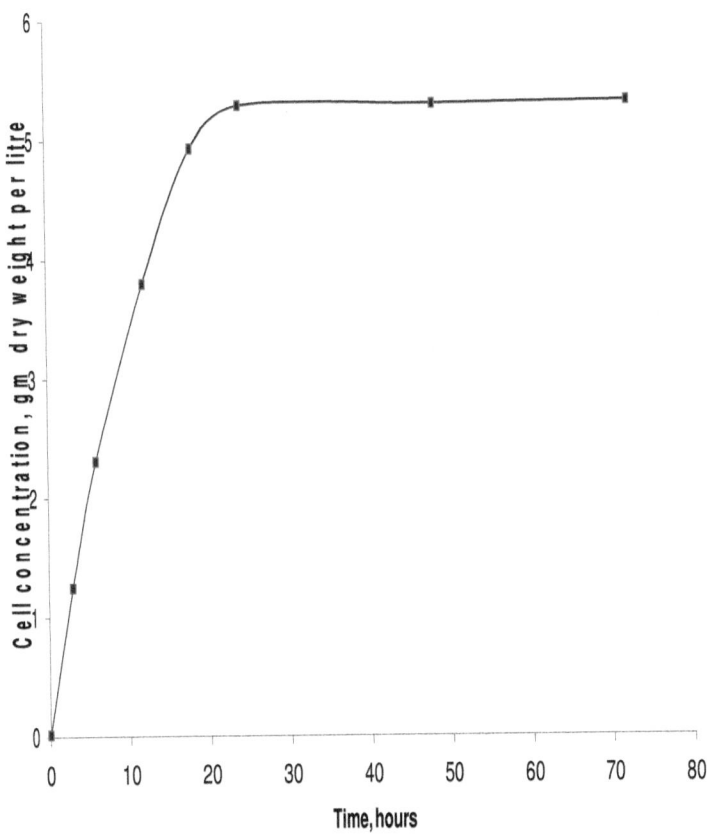

**Figure 7.5:** Biomass cell concentration for basal medium with 2.5% corn starch

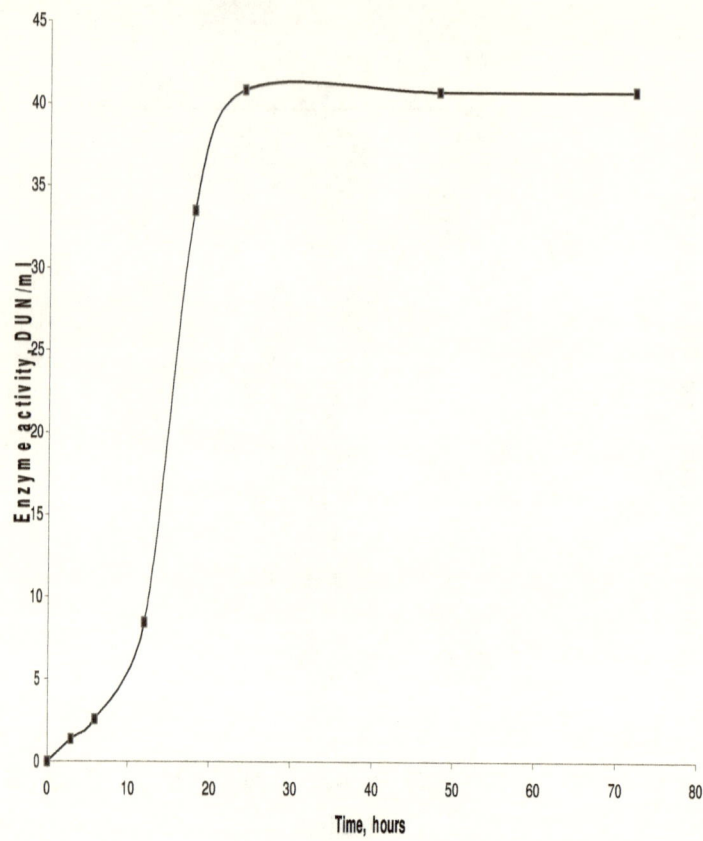

**Figure 7.6:** Enzyme activity for basal medium with 2.5% corn starch concentration

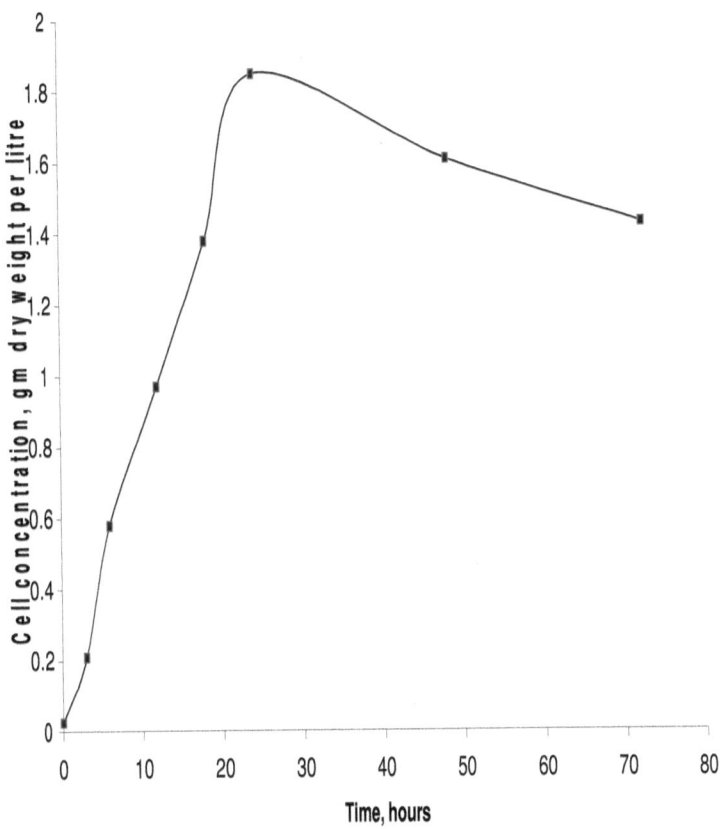

**Figure 7.7:** Biomass cell concentration for Basal medium with 3% defatted Cotton seed

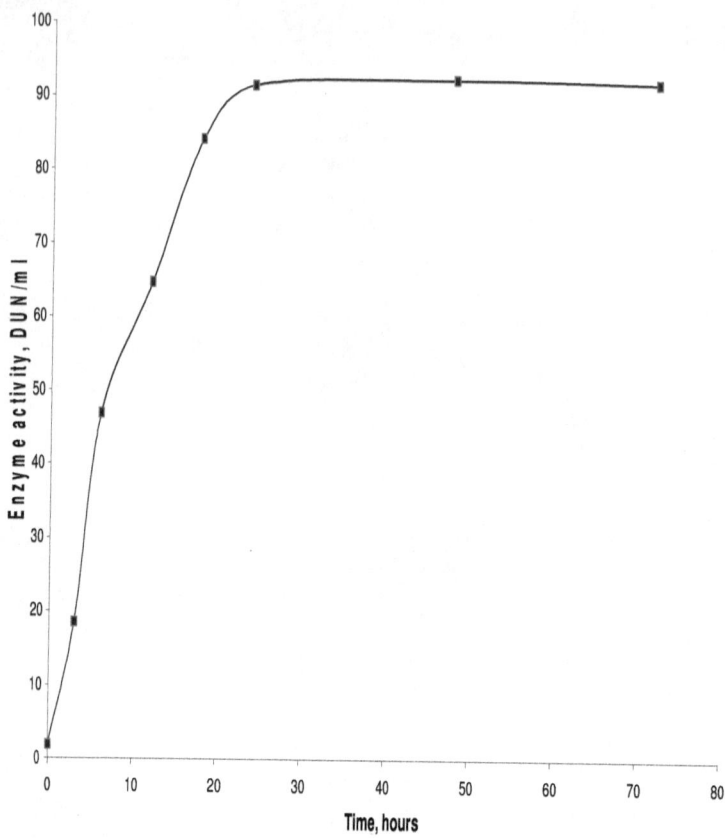

**Figure 7.8:** Enzyme activity for Basal medium with 3% defatted Cotton seed concentration

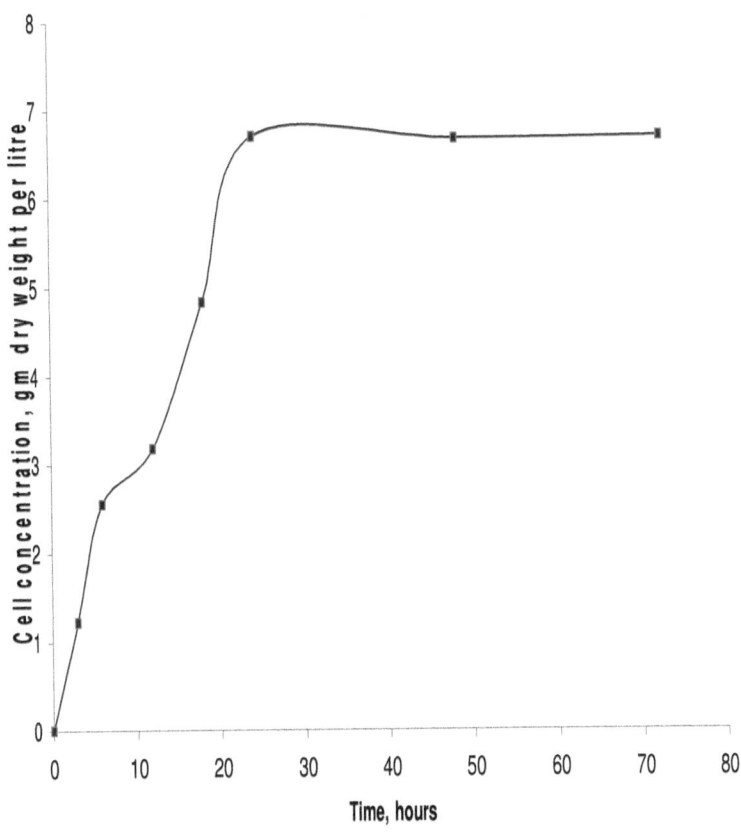

**Figure 7.9:** Biomass cell concentration for basal medium with 0.5% corn starch

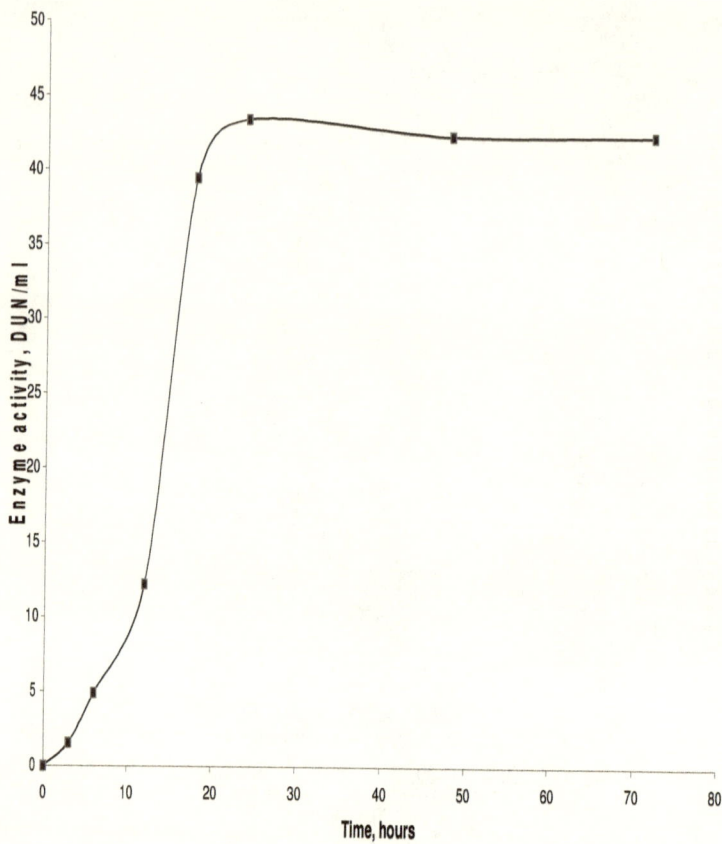

**Figure 7.10:** Enzyme activity for basal medium with 0.5% corn starch concentration

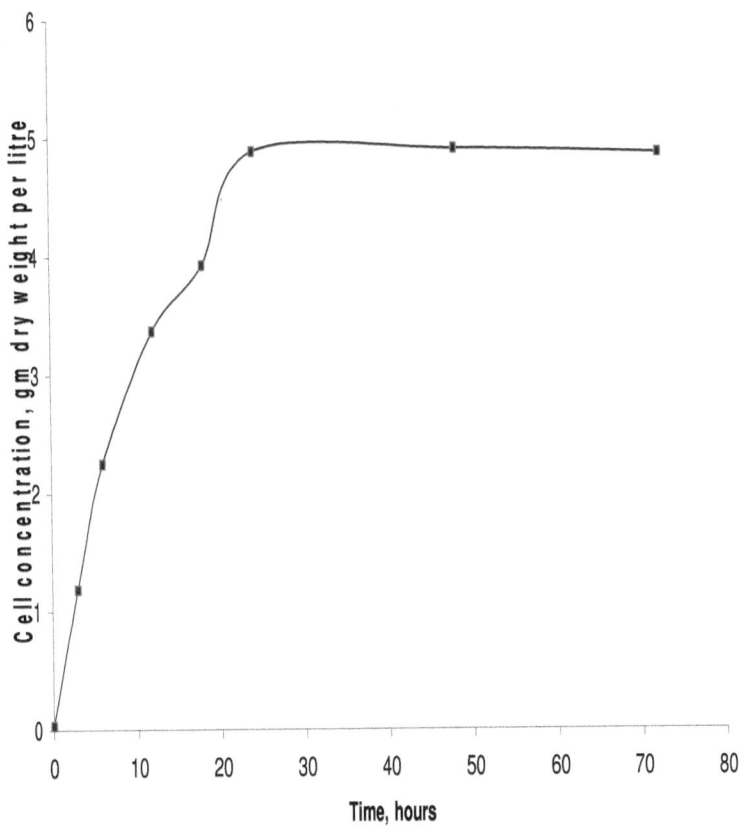

**Figure 7.11:** Biomass cell concentration for Basal medium with 2% Mustard seed

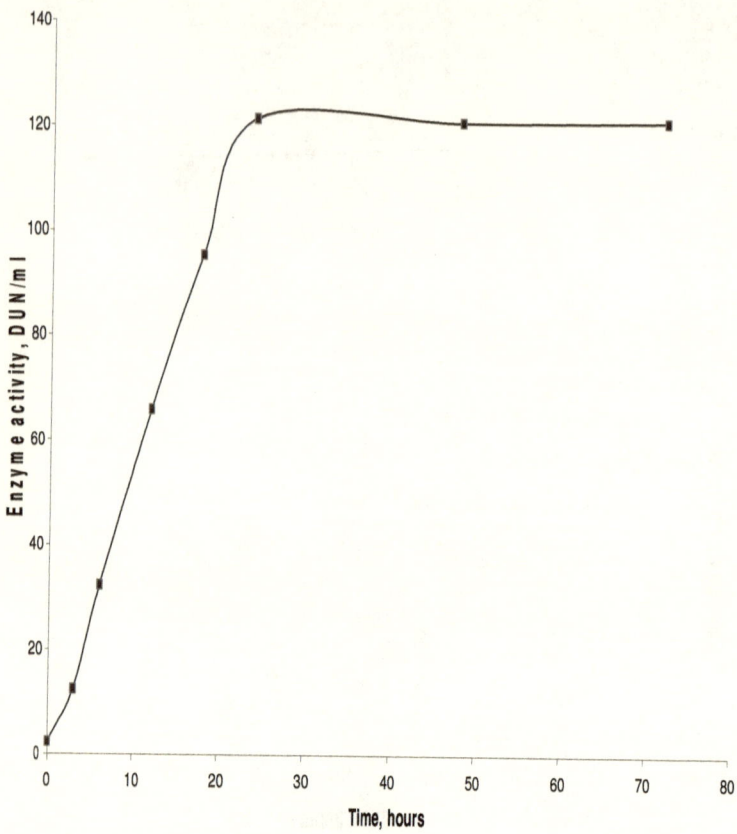

**Figure 7.12:** Enzyme activity for Basal medium with 2% Mustard seed concentration

# **CONCLUSION**

The Bacterial strain, *Bacillus lichenoformis* NCIM 2051 was obtained from National Chemical Laboratory, Pune which produced high temperature alkaline α – amylase. The optimum cultural conditions are found to be 35°C, pH 7 and 300 rpm. The α – amylase produced from this Bacterial strain, *Bacillus lichenoformis* was quite active even at 100°C, however it showed optimum activity at 90°C, and also it exhibited optimum activity in the broad pH range 5.5 – 10, thus α- amylase of *Bacillus lichenoformis* seems to have a very broad pH range. A low cost synthetic medium producing large quantities of α – amylase has been developed from *bacillus lichenoformis* was used for α – amylase production. The α – amylase of this strain showed excellent stability at high temperatures and over a wide pH range. The cell mass concentration and the enzyme activity were determined and optimized. The low cost medium which contains, Defatted soya flour, Defatted cottonseed, and Mustard seed, produces two times more enzyme than the high cost synthetic medium (using yeast extract and peptone) in the B. Braun Biostat E fermentor. So it is further suggested to change the cheapest nitrogen sources in this low cost medium like corn steep liquor.

# BIBLIOGRAPHY

1. Agrawal M., Pradeep S., Chandraraj K., Gummadi S.N. Hydrolysis of starch by amylase from *Bacillus* sp. KCA102: a statistical approach. Process Biochemistry. 2005; 40: 2499–2507.

2. Aguilar G., Morlon-Guyot J., Trejo-Aguilar B., Guyot J.P. Purification and Characterization of an extracellular alpha-amylase produced by *Lactobacillus manihotivorans* LMG 18010(T), an amylolytic lactic acid bacterium. Enzyme Microb Technol. 2000; 27: 406–413.

3. Ahlawat S., Dhiman S.S., Battan B., Mandhan R.P., Sharma J. Pectinase production by *Bacillus subtilis* and its potential application in bio preparation of cotton and micropoly fabric. Process Biochemistry. 2009; 44: 521–526.

4. Amoozegar M.A., Malekzadeh F., Malik K.A. Production of amylase by newly isolated moderate halophile, *Halobacillus* sp. strain MA-2. J Microbial Methods. 2003; 52: 353–359.

5. Arauza L.J., Jozalaa A.F., Mazzolab P.G., Penna T.C.V. Nisin biotechnological production and application: a review. Trends Food Sci Technol. 2009; 20: 146–154.

6. Asgher M., Asad M.J., Rahman S.U., Legge R.L. A thermostable α-amylase from a moderately thermophilic *Bacillus subtilis* strain for starch processing. J Food Process Eng. 2007; 79: 950–955.

7. Asoodeh A., Chamani J., Lagzian M. A novel thermostable, acidophilic alpha-amylase from a new thermophilic "*Bacillus* sp. Ferdowsicous" isolated from Ferdows hot mineral spring in Iran: Purification and biochemical characterization. Int J Biol Macromol. 2010; 46: 289–297.

8. Baysal Z., Uyar F., Aytekin C. Solid state fermentation for production of α-amylase by a thermo tolerant *Bacillus subtilis* from hot-spring water. Process Biochemistry. 2003; 38: 1665–1668.

9. Brayer G.D., Luo Y., Withers S.G. The structure of human pancreatic alpha-amylase at 1.8 A resolution and comparisons with related enzymes. Protein Sci. 1995; 4: 1730 – 1742.

10. Bruinenberg P.M., Hulst A.C., Faber A., Voogd R.H. A process for surface sizing or coating of paper. In: European Patent Application. 1996

11. Calderon M., Loiseau G., Guyot J.P. Fermentation by *Lactobacillus fermentum* Ogi E1 of different combinations of carbohydrates occurring naturally in cereals: consequences on growth energetics and alpha-amylase production. Int J Food Microbiol. 2003; 80: 161–169.

12. Chen W.M., Chang J.S., Chiu C.H., Chang S.C., Chen W.C., Jiang C.M. *Caldimonas taiwanensis* sp. nov., α-amylase producing bacterium isolated from a hot spring. Syst Appl Microbiol. 2005; 28: 415–420.

13. Chi M., Chen Y., Wu T., Lo H., Lin L. Engineering of a truncated α-amylase of *Bacillus* sp. strain TS-23 for the simultaneous improvement of thermal and oxidative stabilities. J. Biosci. Bioeng. 2009;xx:xxx–xxx.

14. Chi Z., Chi Z., Liu G., Wang F., Ju L., Zhang T. *Saccharomycopsis fibuligera* and its applications in biotechnology. Biotechnology Adv. 2009; 27: 423–431.

15. Coronado M., Vargas C., Hofemeister J., Ventosa A., Nieto J.J. Production and biochemical characterization of an alpha-amylase from the moderate halophile *Halomonas meridiana*. FEMS Microbiology Lett. 2000; 183: 67–71.

16. Couto S.R., Sanromán M.A. Application of solid-state fermentation to food industry- A review. Journal of Food Engineering. 2006; 76: 291–302.

17. De Moraes L.M., Astolfi-Filho S., Oliver S.G. Development of yeast strains for the efficient utilisation of starch: evaluation of constructs that express alpha-amylase and glucoamylase separately or as bi functional fusion proteins. Appl Microbiol Biotechnol. 1995; 43: 1067–1076.

18. Deutch C.E. Characterization of a salt-tolerant extracellular α-amylase from *Bacillus dipsosauri*. Lett Appl Microbiol. 2002; 35: 78–84.

19. Djekrif-Dakhmouche S., Gheribi-Aoulmi Z., Meraihi Z., Bennamoun L. Application of a statistical design to the optimization of culture medium for α-amylase production by *Aspergillus niger* ATCC16404 grown on orange waste powder. J Food Process Eng. 2006; 73: 190–197.

20. Feitkenhauer H. Anaerobic digestion of desizing wastewater: influence of pretreatment and anionic surfactant on degradation and intermediate accumulation. Enzyme Microb. Technol. 2003; 33: 250–258.

21. Francis F., Sabu A., Nampoothiri K.M., Ramachandran S., Ghosh S., Szakacs G., Pandey A. Use of response surface methodology for optimizing process parameters for the production of α-amylase by *Aspergillus oryzae*. Biochem. Eng. J. 2003;15: 107–115.

22. Gangadharan D., Sivaramakrishnan S., Nampoothiri K.M., Sukumaran R.K., Pandey A. Response surface methodology for the optimization of alpha amylase production by *Bacillus amyloliquefaciens*. Bioresour Technol. 2008; 99: 4597–4602.

23. Gavrilescu M., Chisti Y. Biotechnology-a sustainable alternative for chemical industry. Biotechnol Adv. 2005; 23: 471–499.

24. Ghorai S., Banik S.P., Verma D., Chowdhury S., Mukherjee S., Khowala S. Fungal biotechnology in food and feed processing. Food Res. Int. 2009; 42: 577–587.

25. Glymph J.L., Stutzenberger F.J. Production, purification, and characterization of alpha-amylase from Thermomonospora curvata. Appl Environ Microbiol. 1977;34: 391–397.

26. Gomes I., Gomes J., Steiner W. Highly thermostable amylase and pullulanase of the extreme thermophilic eubacterium *Rhodothermus marinus*: production and partial characterization. Bioresour Technol. 2003;90: 207–214.

27. Goto C.E., Barbosa E.P., Kistner L.C., Moreira F.G., Lenartovicz V., Peralta R.M. Production of amylase by *Aspergillus fumigatus* utilizing alpha-methyl-D-glycoside, a synthetic analogue of maltose, as substrate. FEMS Microbiol Lett. 1998;167: 139–143.

28. Goyal N., Gupta J.K., Soni S.K. A novel raw starch digesting thermostable α-amylase from *Bacillus* sp. I-3 and its use in the direct hydrolysis of raw potato starch. Enzyme Microb. Technol. 2005; 37:723–734.

29. Gupta R., Gigras P., Mohapatra H., Goswami V.K., Chauhan B. Microbial α-amylases: a biotechnological perspective. Process Biochem. 2003; 38: 1599–1616.

30. H.J. Rehm and G. Read, Biotechnology, Volume 7a, Enzyme technology, VCH publishers (1987).

31. Hamilton L.M., Kelly C.T., Fogarty W.M. Production and properties of the raw starch-digesting α-amylase of *Bacillus* sp. IMD 435. Process Biochem. 1999;35: 27–31.

32. Hamilton L.M., Kelly C.T., Fogarty W.M. Purification and properties of the raw starch degrading -amylase of *Bacillus* sp. IMD434. Biotechnol. Lett. 1999;21: 111–115.

33. Hernández M.S., Rodríguez M.R., Guerra N.P., Rosés R.P. Amylase production by *Aspergillus niger* in submerged cultivation on two wastes from food industries. J Food Process Eng. 2006; 73: 93–100.

34. Hmidet N., El-Hadj Ali N., Haddar A., Kanoun S., Alya S., Nasri M. Alkaline proteases and thermostable α-amylase co-produced by *Bacillus licheniformis* NH1: Characterization and potential application as detergent additive. Biochemical Engineering Journal. 2009;47:71–79.

35. Hutcheon G.W., Vasisht N., Bolhuis A. Characterisation of a highly stable alpha-amylase from the halophilic archaeon *Haloarcula hispanica*. Extremophiles. 2005;9: 487–495.

36. Ikram ul H., Ashraf H., Iqbal J., Qadeer M.A. Production of alpha amylase by *Bacillus licheniformis* using an economical medium. Bioresour Technol. 2003; 87:57–61.

37. Iulek J., Franco O.L., Silva M., Slivinski C.T., Bloch C., Jr., Rigden D.J., Grossi de Sa M.F. Purification, biochemical characterization and partial primary structure of a new alpha-amylase inhibitor from *Secale cereale* (rye) Int J Biochem Cell Biol. 2000;32:1195–1204.

38. J. Jayaraman, "Laboratory manual in Biochemistry", (1981).

39. James E. Bailey and David F.Ollis, "Biochemical engineering fundamentals", McGraw.Hill international editions, second edition, (1986).

40. Jaspreet Singha J., Kaurb L., McCarthy O.J. Factors influencing the physico-chemical, morphological, thermal and rheological properties of some chemically modified starches for food applications—A review. Food Hydrocolloids. 2007; 21:1–22.

41. Jensen B., Nebelong P., Olsen J., Reeslev M. Enzyme production in continuous cultivation by the thermophilic fungus, *Thermomyces lanuginosus*. Biotechnology Letters. 2002; 24:41–45.

42. Jin B., van Leeuwen H.J., Patel B., Yu Q. Utilisation of starch processing wastewater for production of microbial biomass protein and fungal α-amylase by *Aspergillus oryzae*. Bioresour. Technol. 1998; 66:201–206.

43. Kajiwara Y., Takbshima N., Ohba H., Omori T., Shimoda M., Wada H. Production of Acid-Stable α-Amylase by *Aspergillus* during Barley *Shochu-Koji* Production. J. Ferment. Bioeng. 1997; 84:224–227.

44. Kammoun R., Naili B., Bejar S. Application of a statistical design to the optimization of parameters and culture medium for alpha-amylase production by Aspergillus oryzae CBS

819.72 grown on gruel (wheat grinding by-product) Bio resources Technol. 2008;99:5602–5609.

45. Kandra L. α-Amylases of medical and industrial importance. Journal of Molecular Structure (Theochem) 2003:666–667. 487–498.

46. Kathiresan K., Manivannan S. α-Amylase production by *Penicillium fellutanum* isolated from mangrove rhizosphere soil. Afr. J. Biotechnol. 2006;5: 829–832.

47. Khoo S.L., Amirul A.A., Kamaruzaman M., Nazalan N., Azizan M.N. Purification and characterization of alpha-amylase from Aspergillus flavus. Folia Microbiol (Praha) 1994; 39: 392–398.

48. Kirk O., Borchert T.V., Fuglsang C.C. Industrial enzyme applications. Curr Opin Biotechnol. 2002; 13: 345–351.

## I want morebooks!

Buy your books fast and straightforward online - at one of the world's fastest growing online book stores! Environmentally sound due to Print-on-Demand technologies.

Buy your books online at
# www.get-morebooks.com

Kaufen Sie Ihre Bücher schnell und unkompliziert online – auf einer der am schnellsten wachsenden Buchhandelsplattformen weltweit! Dank Print-On-Demand umwelt- und ressourcenschonend produziert.

Bücher schneller online kaufen
# www.morebooks.de

OmniScriptum Marketing DEU GmbH
Heinrich-Böcking-Str. 6-8
D - 66121 Saarbrücken
Telefax: +49 681 93 81 567-9

info@omniscriptum.com
www.omniscriptum.com

www.ingramcontent.com/pod-product-compliance
Lightning Source LLC
Chambersburg PA
CBHW031536210526
45464CB00003B/1035